RAND PROJECT AIR FORCE

T0108993

From Theory to Practice

People's Liberation Army Air Force Aviation
Training at the Operational Unit

Lyle J. Morris, Eric Heginbotham

Prepared for the United States Air Force

For more information on this publication, visit www.rand.org/t/RR1415

Library of Congress Cataloging-in-Publication Data is available for this publication.
ISBN: 978-0-8330-9497-1

Published by the RAND Corporation, Santa Monica, Calif.
© Copyright 2016 RAND Corporation
RAND® is a registered trademark.

Support RAND
Make a tax-deductible charitable contribution at
www.rand.org/giving/contribute

www.rand.org

Preface

About the China Aerospace Studies Institute

The China Aerospace Studies Institute (CASI) was created in 2014 at the initiative of the Headquarters, U.S. Air Force. CASI is part of the RAND Corporation's Project AIR FORCE (PAF); Air University and Headquarters Pacific Air Forces are key stakeholders. The purpose of CASI is to advance understanding of the capabilities, operating concepts, and limitations of China's aerospace forces. Its research focuses on the People's Liberation Army Air Force (PLAAF), Naval Aviation, PLA Rocket Forces, and the Chinese military's use of space. CASI aims to fill a niche in the China research community by providing high-quality, unclassified research on Chinese aerospace developments in the context of U.S. strategic imperatives in the Asia-Pacific region. CASI will transition to an independent, Air Force–supported organization in fiscal year 2017, with personnel at Ft. McNair in Washington, D.C., and at Air University, Maxwell Air Force Base, Alabama.

More than two decades ago, the armed forces of the People's Republic of China embarked on a series of efforts to transform from a massive, low-tech force focused on territorial defense to a leaner, high-tech force capable of projecting power and influence around the peripheries of the nation, even against the expeditionary forces of the most capable adversary. By developing operational strategies tailored to meeting this objective; reforming the organization, training, and doctrine of the armed forces; and, perhaps most importantly, making large and sustained investments in new classes of weapon systems, China may be on the cusp of realizing the ambitious goals it has laid out for its armed forces. Without doubt, China's air and maritime forces today field capabilities that are compelling U.S. military planners to rethink their approach to power projection and to reorient important components of their modernization programs. China, in short, has become the "pacing threat" for the U.S. Air Force and Navy. As such, the importance of understanding the capabilities and limitations of China's air, naval, and space forces is self-evident. The mission of CASI is to contribute to that understanding.

CASI's research team brings to this work a mastery of research methods, understanding of China's military capabilities and doctrine, and the ability to read and understand Chinese writings. When undertaking research for CASI reports, analysts used a variety of Chinese-language primary-source documents on PLA Army and PLAAF training, operations and doctrine. This includes *Kongjun Bao* (Air Force News) and *Huojianbing Bao* (Rocket Force News)—the daily newspapers of the PLAAF and PLA strategic missile forces—as well as defense white papers, PLA encyclopedias, and books by military officers and academics affiliated with the PLA (such as the Academy of Military Science). These publications are considered authoritative assessments and reporting on training, strategy, and concepts for how the PLAAF and missile forces prepare for military operations and warfare in general. It is important to acknowledge, however, that these PLA publications also have some weaknesses, and that reliance on open sources necessarily has some limitations. The growing availability of primary-source material helps to compensate for at least some of these challenges.

Additional information about CASI is available on RAND's CASI website: www.rand.org/paf/casi

RAND Project AIR FORCE

RAND Project AIR FORCE (PAF), a division of the RAND Corporation, is the U.S. Air Force's federally funded research and development center for studies and analyses. PAF provides the Air Force with independent analyses of policy alternatives affecting the development, employment, combat readiness, and support of current and future air, space, and cyber forces. Research is conducted in four programs: Force Modernization and Employment; Manpower, Personnel, and Training; Resource Management; and Strategy and Doctrine. The research reported here was prepared under contract FA7014-06-C-0001.

Additional information about PAF is available on our website: www.rand.org/paf

Table of Contents

Figures

Summary

The People's Liberation Army Air Force (PLAAF) has embarked on a major institutional reform to train and equip a modern, professional aviator corps. At the heart of this reform is an effort to train officers under what the People's Liberation Army refers to as "actual combat conditions" (实战条件). Such an emphasis reflects an acknowledgment by senior leaders that the PLAAF, and by extension the entire Chinese armed forces, is ill prepared to "fight and win" wars against potentially superior military competitors and must therefore reinvigorate training programs to meet the missions that the Chinese Air Force may be called on to execute.

This report seeks to assess PLAAF pilot proficiency by examining training held at operational aviation units in the context of the larger PLAAF training system. The first section analyzes the hierarchy of PLAAF training guidance. The second section examines the PLAAF training system for pilots prior to their arrival at their operational units, to include an examination of the theoretical and practical methods of instruction. The third section covers the transition to the PLAAF operational aviation unit and the content of unit training, including an analysis of the PLAAF annual training cycle. The fourth section addresses the development and training of instructor pilots at the unit level. The final section seeks to evaluate the operational competency and weaknesses within the PLAAF aviation training system.

Acknowledgments

The authors would like to acknowledge the support of RAND Project AIR FORCE and Carl Rehberg, Lt Col Benjamin Carroll, and the staff in the Strategy Division, DCS Strategic Plans and Requirements, Headquarters U.S. Air Force. We also thank the valuable insights provided by Maj David F. John and others who wish to remain anonymous but whose contributions were critical to our work.

Abbreviations

AAA	anti-aircraft artillery
AWS	U.S. Air Force Weapons School
CASI	Chinese Aerospace Studies Institute
CEME	complex electromagnetic environments
FITB	Flight Instructor Training Base
FTU	Formal Training Unit
GSD	General Staff Department
IFF	Introduction to Fighter Fundamentals
MRAF	Nanjing Military Region Air Force
OTS	Officer Training School
PAF	Project AIR FORCE
PIT	U.S. Air Force Pilot Instructor Training
PLA	People's Liberation Army
PLAAF	People's Liberation Army Air Force
PLAAF AU	People's Liberation Army Air Force Aviation University
SAM	surface-to-air missiles
SoS	system of systems
UPT	Undergraduate Pilot Training
USAF	U.S. Air Force

1. Introduction

An examination of People's Liberation Army Air Force (PLAAF) fighter and attack pilot training at the unit level can assist in the evaluation of PLAAF training as a whole.[1] We exploit open-source literature—to include PLAAF service newspaper *Kongjun Bao* (Air Force News)—to produce an overview of the duration, nature, and content of fighter pilot training. Where appropriate and possible, we compare this training with unit-level pilot training in the U.S. Air Force (USAF).

The extent to which Chinese efforts to improve the realism and rigor of training at the unit level succeed ultimately depends on institutional factors. Our findings suggest that the PLAAF is professionalizing unit training through adherence to less-scripted, combat-realistic training that trains for the battlefield, not for the test. The PLAAF seeks to cultivate more autonomous decisionmaking among fighter pilots and has recently begun to shift training away from an emphasis on ground control to a system that encourages independent decisionmaking. However, structural weaknesses persist, and aviators continue to struggle under less-scripted environments and when coordinating operations with other elements. The continued reliance on conscripts for a significant portion of the enlisted force keeps training tied to an annual training cycle and prevents units from maintaining year-round readiness. At the same time, the PLAAF lacks an analogue to the USAF Weapons School (AWS) or any other similar means of developing the combat skills of midcareer pilots who might then serve as training officers at the unit level. The lack of an AWS equivalent also limits Chinese ability to maintain broad uniformity in tactical practices. Unit commanders and other senior cadre maintain a level of autonomy in instituting the content and scope of certain training schedules, including the formulation of training tasks and practices. But underdeveloped linkages between unit training and PLAAF headquarters, as well as between the units and PLAAF flight colleges, continue to hinder the institutionalization of training within the PLAAF.

Before analyzing the content of PLAAF pilot raining at the unit level, we outline the system of training guidance that dictates the types of pilot training activities that PLAAF units undertake.

---

[1] Foundational assessments of PLAAF modernization efforts include Richard P. Hallion, Roger Cliff, and Phillip C. Saunders, *The Chinese Air Force: Evolving Concepts, Roles, and Capabilities*, Fort McNair, Washington, D.C.: National Defense University, 2012; National Air and Space Intelligence Center, *People's Liberation Army Air Force 2010*, Wright-Patterson Air Force Base, Ohio, National Air and Space Intelligence Center: August 1, 2010; and Roger Cliff, John F. Fei, Jeff Hagen, Elizabeth Hague, Eric Heginbotham, and John Stillion, *Shaking the Heavens and Splitting the Earth: Chinese Air Force Employment Concepts in the 21st Century*, Santa Monica, Calif.: RAND Corporation, MG-915-AF, 2010.

2. PLAAF Training Guidance

Training guidance for the People's Liberation Army (PLA) follows a highly centralized, top-down approach. All services and branches within the PLA, including the PLAAF, are subject to "training guidance" issued by the General Staff Department (GSD). There are three levels of guidance: (1) *Outline of Military Training and Evaluation* (军事训练与考核大纲); (2) *Annual Directive on Military Training;* and (3) *Annual Training Cycle.* The top-down nature of military guidance in the Chinese system produces common priorities, lexicon, and concepts throughout the PLA's various services and branches but has not produced a high degree of "jointness." Developments in regard to joint training are discussed in a separate paper in this series.

The *Outline of Military Training and Evaluation,* hereafter referred to as the Outlines, is typically issued roughly once in a decade and is the authoritative guide to all services within the PLA on how they should carry out training in the coming years.[2] The Outlines represent a major effort among the various general departments and Central Military Commission and often undergo years of rigorous study and revision before the GSD officially releases them.[3] Before 1995, most of the armed services issued their own guidelines based on the parameters described in the Outlines. Recent Outlines, however, explicitly describe service- and branch-specific guidelines and regulations to be carried out. New Outlines typically have one theme and various "training areas" designed to improve overall PLA capabilities and operations, although some Outlines include more specificity, such as stipulating flight-time quotas for certain training

---

[2] Among other items, the Outlines include training goals, principles, content, implementation phases and procedures, timing, methods, and quality-control inspection procedures for all the armed services within China. See National Air and Space Intelligence Center, 2010, p. 61.

[3] For example, the 2015 Outline was released on January 29, 2015, after undergoing two years of "compilation and revision" based on feedback from the various military departments. See Liang Pengfei and Li Yuming, "Joint Examination on Trial Training Outline of All-Army New Training Program Completed," *China Military Online*, February 5, 2015. The 2009 Outline reportedly involved input from the four general departments, all four services of the PLA (army, navy, air force, and Second Artillery), and "over 1,000 military training specialists from institutions, troops, military academies and schools, and scientific research institutes." See Liu Fengan and Wu Tianmin, "New-Generation 'Outline of Military Training and Evaluation' Promulgated" ["新一代《军事训练与考核大纲》颁发"], *Jiefangjun Bao*, July 25, 2008, p. 1.

subjects.[4] The Outlines typically undergo a year of trials before they are officially put into use.[5] Since 1957, eight Outlines have been released.

The second training guideline is the *Annual Directive on Military Training* (新年度全军军事训练任务), hereafter referred to as Directives. As opposed to the Outlines, which are issued roughly every decade and set the overall course of training objectives, Directives are issued annually at the beginning of every calendar year after the completion of the "All-Army Training Conference" (全军训练会议). Navy, Air Force, Second Artillery, and each military region headquarters then hold meetings after this conference to discuss and promulgate service-specific "training plans" (年度训练计划) for the coming year. Each subordinate unit then holds meetings concerning the guidance. The PLAAF daily newspaper, *Kongjun Bao*,[6] reported on the process of an unidentified Shenyang Air Regiment discussing the "plan" for the upcoming year:

> In the course of formulating this year's training plan, the regiment considered issues and planned force building in accordance with the military strengthening goal. . . . In this year's training plan, the regiment focuses on core capabilities training, mainly honing the force's defense penetration and assault skills and enhancing the actual fighting capabilities of moving, penetrating, searching, attacking.[7]

Directives build off the broad parameters presented in the Outlines, stipulating the plans, training subjects, and conditions under which the various services and branches are to carry out training for that year. The 2015 Directive, for example, calls for the PLA to "continue to grasp

---

[4] For example, one *Kongjun Bao* article noted that a certain air unit "sustained very low altitude flight time that reached the upper limit mandated in the Outline of Military Training and Evaluation." See Yan Guoyou and Liao Qirong, "PLA Nanjing MRAF Division: Combat Realistic Training Routinized" ["南空航空兵某师常态化开展实战化训练"], *Kongjun Bao,* August 15, 2013.

[5] For example, the 2008 Outline highlighted joint operations as its main theme. See Liu Fengan and Wu Tianmin, 2008, p. 1. One report on the 2015 Outline noted that it was "completed on January 29, 2015" and that "in 2015, trial training and demonstration of the new Outline will be carried out in various military regions, arms and services and the Armed Police Force of the entire armed forces." See Liang Pengfei and Li Yuming, 2015.

[6] Published five days a week, *Kongjun Bao* (空军报) is the PLAAF's only service-specific newspaper, featuring articles on Chinese Communist Party guidelines, officer biographies, and most importantly, detailed descriptions of training events that occur throughout the year. It is the most authoritative open-source Chinese-language resource on PLAAF training available to analysts.

[7] Wang Zhonghai and Liu Daquan, "Air Regiment of the Shenyang Military Region Air Force Energetically Conducts Core Capabilities Training, Hones Defense Penetration and Assault Skills," *Kongjun Bao*, February 10, 2014.

closely realistic combat training" (继续抓好实战化军事训练)—a consistent theme found in previous Directives and Outlines—by focusing on the following points of emphasis:[8]

- nighttime battle training (夜战)
- complex electromagnetic conditions (复杂电磁环境)
- special geographical environments (特殊地理环境)
- extreme weather conditions (极端天候条).

Finally, these Outlines, Directives, and training plans feed into the PLAAF *Annual Training Cycle* (新年度开训), hereafter referred to as the Cycles, which set the agenda for the upcoming year of training subjects by announcing themes of study.[9] During a PLAAF Party Committee meeting in early January 2014, for example, PLAAF commander-in-chief General Ma Xiaotian referenced the new annual training cycle by urging party members and commanders to "make a good start for flight training in the new annual cycle by strengthening training safety work in main units and in important periods" and directed the PLAAF Party Committee to "send four working groups to various air units to inspect and direct work in the new year."[10] The deployment of small working groups that oversee training execution at the unit level is a new development and speaks to the recent PLAAF reforms that emphasize the institutionalization of training. Subsequent sections of the report will address the different components of the Cycle in more detail.

Despite these top-down guidelines, aviation units do appear to have a certain degree of autonomy in setting and adjusting training subjects. One August 2013 *Kongjun Bao* article, for example, reported that a Nanjing Military Region Air Force (MRAF) division developed numerous new defense penetration, assault, and long-range precision strike combat methods "in response to needs that arose during the course of training and drilling."[11] Thus, aviation unit commanders appear to adjust and tailor certain types of training activities in accordance with issues or challenges identified throughout a training cycle.

---

[8] Liang Pengfei, "PLA General Staff Department Promulgates Directive on 2015 PLA Military Training," ["总参部署新年度全军军事训练任务"], *Jiefangjun Bao*, January 18, 2015.

[9] Themes vary for every cycle, but focus on conducting training that mimics actual warfighting conditions. For example, the theme of the 2013 cycle called for "being able to defend, penetrate, and attack triumphantly . . . while adhering closely to actual operations." See Xu Weigang and Si Han, "An Air Brigade of the Nanjing Military Region Air Force Organizes Highly Demanding Training in Accordance with Actual Operation Requirements" ["紧贴实战应用组织高难课目训练"], *Kongjun Bao,* June 27, 2013.

[10] Tian Wei, "Solidly Do a Good Job in Conducting Flight Training in the New Annual Cycle with Concentrated Attention and Rigorous Organization" ["集中经历严密组织切实抓好新年度飞行训练"], *Kongjun Bao*, January 8, 2014.

[11] Yan Guoyou and Liao Qirong, 2013.

4

As the subsequent sections of this report will highlight, recent PLAAF unit training incorporates many of the above-mentioned objectives and themes into training lessons and exercises.

3. PLAAF Aviation Pre-Unit Training System

Similar to USAF pilots, Chinese pilots must go through a rigorous process of theoretical study at a training academic institution (for the PLAAF, this occurs at the PLAAF Aviation University [PLAAF AU] [空军航空大学] in Changchun, Jilin Province) before transitioning to the practical application (实际操练) phase of training.[12] A brief analysis of each of the three phases—academy learning and basic training, flight college training, and flight transition training—is presented below.

Phase 1: Academic Institution Learning and Basic Training (Four Years)

The most common channel for aspiring pilots in China is to first enter the PLAAF AU out of high school and begin learning basic aviation theory and navigation skills.[13] Based on content outlined in the *PLAAF Encyclopedia*, aviation theory study covers, broadly speaking, the following categories:[14]

- *foundational theory*, which encompasses aviation theory, aeronautics propulsion, airframe theory, air flight theory, air-launched missile theory, weather patterns, aviation electronic theory, aviation ordnance theory, and fire control system theory
- *applied theory*, which encompasses navigation, assault, bombing, surveillance, and other applied methods of theory
- *combat theory*, which includes military thought, combat theory, air campaigns, tactics, combat maneuvers, and flight accompaniment tasks and methods.

After a foundation in aviation theory is completed, pilots begin to learn applied concepts of aviation by holding flight simulation training (飞行模拟训练).[15] Flight simulation training

---

[12] *China Air Force Encyclopedia*, Vol. 1, Yao Wei, ed., Beijing, China: Aviation Industry Press, November 2005, p. 274.

[13] This scenario represents cadets coming out of high school and does not include paths from other programs such as the National Defense Student Program or graduates selected from other PLA academic institutions. The four-phased training program outlined here only began at the end of 2011 and thus does not reflect other pilots who are already assigned to operational units prior to 2011. For more on these "phases," see Kenneth W. Allen, "PLA Air Force, Naval Aviation and Army Aviation Aviator Recruitment, Education, and Training," Washington, D.C.: The Jamestown Foundation, February 2016.

[14] *China Air Force Encyclopedia*, 2005, p. 274.

[15] *China Air Force Encyclopedia*, 2005, pp. 274–275.

involves the use of a ground-operated simulator to practice basic flying skills such as takeoffs, landings, and inflight emergency procedures. Once pilots become comfortable with the skills needed to fly on a simulator, they begin conducing actual flights for about six months in a trainer aircraft called a CJ-6 (Figure 3.1), in which pilots hone basic aviation skills associated with operating an aircraft, called *aviation navigation skills training* (飞行驾驶技术训练). Skills training involves basic concepts such as takeoffs, landings, spins, and emergency procedures; basic navigation; instrument flying; night flying; and formation flying.[16] This first phase typically takes four years to complete and provides theoretical and basic practical aviation training for cadets.

Figure 3.1. CJ-6 Basic Flight Trainer

NOTE: A privately owned CJ-6 aircraft.
SOURCE: Photo by Cialowitz via Wikipedia Commons (CC BY-SA 3.0).

Phase 2: Flight College Training (One to Two Years)

Upon completion of basic training, cadets enter the second phase of training, known as the *advanced training stage*, which occurs at one of the PLAAF's three flight colleges (飞行学院). During this stage, pilots gain practical hands-on experience with a more advanced trainer, which in most cases is the K-8 (Figure 3.2). Training consists of ground-based training, flying with a flight instructor, and solo flight training and includes such skills as all-weather training,

---

[16] *China Air Force Encyclopedia*, 2005, p. 276. For more information on the different stages and content of PLAAF aviation academy training, see Allen, 2016.

formation flying, simple aerobatics, and navigation. In order to move on to each successive training task, students must pass the necessary examination and exhibit mastery of the skill being learned. Upon completion of training in this phase, cadets have accumulated 150–200 flight hours.[17] It is most likely during this stage of flight training that pilots begin to be assigned to aircraft at the operational units. This process is based on an evaluation of pilot performance and ranking beginning in Phase 1 and continuing into the beginning stages of Phase 2. Which type of aircraft pilots are assigned to is also likely informed by evolving roles and missions of the PLAAF and its force structure requirements. This phase of training lasts one to two years and prepares pilots for the final stage of training, called *flight transition training*.

In February 2015, the PLAAF introduced supersonic aircraft trainers into the advanced flight training phase of flight colleges, which will reportedly "greatly reduce the training cycle of new cadets."[18] As of now, PLAAF pilots do not begin training on supersonic fighter trainers until the third phase, during flight transition training at their operational unit. For comparison, USAF fighter and bomber pilots start training on the supersonic T-38 Talon in the second phase of specialized undergraduate pilot training (Appendix A).

Figure 3.2. K-8 Advanced Flight Trainer Aircraft

NOTE: Pakistani Air Force K-8 trainer aircraft at the 2010 Zhuhai Airshow in China.
SOURCE: Photo by Peng Chen via Flickr (CC BY_SA 2.0).

---

[17] Allen, 2016.

[18] "Aviation Academies Use Supersonic Fighter in Pilot Training" ["超音速歼击机首次进入空军飞行院校"] *China Military Online*, February 28, 2015.

Phase 3: Flight Transition Training Phase (One Year)

Upon graduating from flight college, new pilots deploy to a separate training unit within an operational base, where they are assigned to a transition training flight group for one year. During this phase, pilots train on advanced supersonic trainer aircraft and then begin the process of being "fit" to the type of aircraft they will fly operationally (what the PLAAF refers to as *gaizhuang feixing xunlian*, 改装飞行训练).[19] This period allows pilots to familiarize themselves with the features and advanced instruments of the aircraft as well as develop tactical skills with their assigned aircraft.

According to the *PLAAF Encyclopedia*, combat training (飞行战斗技术训练) during this phase includes weapons and instrument training, air assault, air-to-air combat, air-to-ground targeting, surveillance, and air transport. This is also the stage in which pilots practice more extreme higher- and lower-altitude flying and flying in adverse weather conditions. Pilots also practice basic tactics and combined arms training. By the time a PLAAF pilot is assigned to a flight squadron at an operational base, he or she will have undergone approximately two to three years of practical training as well as roughly four years of academic and theoretical study. Compared with his or her USAF counterpart, Chinese pilots appear to spend more time in flight school and associated aeronautics training before arriving at the operational unit (Appendix A). There is some evidence that the washout rate for aspiring cadets progressing from Phase 1 to assignment to their operational unit is around 50 percent. An article from December 29, 2014, in *China Military Online* states that more than 60 percent of the approximately 1,000 new cadets per year at the PLAAF's Air University are eliminated during their four-year course.[20]

---

[19] *China Air Force Encyclopedia*, 2005, p. 282. Within PLAAF aviation, this can include a fighter, assault, bomber, transport, surveillance, tanker, or airborne warning and control system (AWACS) aircraft.

[20] Allen, 2016.

4. PLAAF Pilot Training at the Unit Level

Arguably the biggest difference between Chinese and U.S. pilot training occurs at the operational unit. Whereas U.S. air units strive for year-round readiness, Chinese pilot training at the unit level revolves around an annual training cycle, which we describe in this chapter. The next chapter treats weaknesses in the system of developing the air combat skills of individuals who will serve as trainers at the unit level, as well as Chinese efforts to remedy that problem.

Aviation Unit Training Cycle

Several factors dictate or shape the annual training cycle associated with PLAAF fighter units. The first is institutional. PLAAF training must accommodate the two-year enlistment terms of PLAAF conscripts, who make up a large proportion of PLAAF service members in maintenance and other ground-support positions. These individuals typically arrive at their designated units in the winter.[21] Although conscripts currently account for a small and decreasing percentage of total personnel, running large, integrated exercises may be difficult until the proficiency of these conscripts is raised to a certain standard. Another driver is the need to adhere to top-down training plans and PLAAF Party Committee directives, which stipulate different types of coordinated evaluations and major exercises to test the ability of pilots to perform under standardized criteria. Because of the size and scale of these exercises and evaluations, they can be carried out only during certain times of the year. The drill and exercise season, for example, typically occurs during the summer.

Estimates vary regarding how much flying time the average PLAAF pilot obtains over the course of an annual training cycle. During a March 2007 visit to the PLAAF's 1st Air Division at Anshan Air Base, then–U.S. chair of the Joint Chiefs of Staff, General Peter Pace, was told that the PLAAF pilots fly an average of 120 hours per year.[22] The division reportedly had 130 pilots and 100 aircraft. In comparison, USAF pilots average 250 flying hours per year with roughly 120 pilots per 100 aircraft.[23] Although these figures are somewhat dated, U.S. pilots almost certainly continue to receive more flight hours than their Chinese counterparts.

---

[21] Bonny Lin and Cristina L. Garafola, *Training the People's Liberation Army Air Force Surface-to-Air Missile (SAM) Forces*, Santa Monica, Calif.: RAND Corporation, RR-1414-AF, 2016.

[22] Jim Garamone, "Pace Visits Chinese Air Base, Checks Out Su-27 Fighter-Bomber," *American Forces Press Service*, March 24, 2007.

[23] Garamone, 2007.

Chinese pilot training at the unit level passes through five partially overlapping segments during the course of the year: new year flight training, training in "subjects" and "topics," peak drills and exercises, a second round of training in "subjects" and "topics," and year-end evaluations (Figure 4.1). The following section summarizes each of these different components of the training cycle.

Figure 4.1. Annual Training Cycle for Notional PLAAF Aviation Base

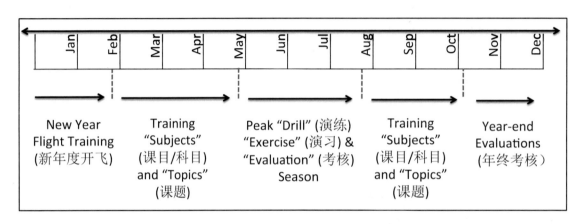

SOURCE: Author's estimate based on *Kongjun Bao* reporting. According to Chinese training regulations, roughly 210 days make up the average training cycle. See *People's Liberation Army Military Training Regulations* (中国人民解放军军事训练条例), Beijing, China: PRC Central Military Commission (中央军委), April 6, 1990.

New Year Flight Training (January–February)

New year flight training (新年度开飞) is the first major training activity of the calendar year and typically lasts into February. Historically, this begins after pilots return from the Chinese Lunar New Year holiday and serves as "refresher" training, testing rudimentary skills and methods and warming up pilots for the remainder of the training cycle. In line with the PLAAF's efforts to adhere more closely to training under "actual combat conditions," however, operational units are increasingly holding new year refresher training during extreme weather events, such as heavy snow and fog. One *Kongjun Bao* article noted how, during one particular training activity in January 2013 at a regiment in Guangzhou, pilots expressed "concern" over the risks of holding difficult training subjects during dense fog and snow conditions that were "just past the minimal takeoff conditions for the warplanes." The pilots recommended that the training be "put off until after the weather had further improved."[24] However, the unit commanders proceeded with

---

[24] Wang Haihua, Zhao Lingyu, and Wang Caishan, "'Airborne Hammer' Strikes Like a Thunderbolt—Eyewitness Account of Strict and Assiduous Organization of Initial Flight Training at Start of the New Year While Closely Adhering to Requirements of Actual Combat by a Certain Regiment of the Guangzhou Military Region Air Force Aviation Corps," *Kongjun Bao*, January 9, 2013.

activities, adding, "if war started today, complex weather situations are nothing to speak of!"[25] Such reports of pilots being challenged to ramp up training of more difficult and dangerous subjects during inclement weather during new year refresher training have been noted in *Kongjun Bao* articles over the last several years.

Subjects and Topics (March–May)

In spring, PLAAF pilots spend the majority of their time on training "subjects" (课目/科目) and training "topics" (课题). Subjects progress from rudimentary to complex tasks and are divided into compulsory (必飞练习), integrated (综合练习), and self-directed (自编飞行练习) elements.[26] Compulsory subjects consist of skills that PLAAF commanders regard as foundational and often mimic the skills that pilots learn in the flight transition training bases, such as basic navigation, flying in low and high altitude, flying in adverse weather conditions, flying at night in normal and adverse weather conditions, and flying in formation. Integrated subjects incorporate more complex subjects, such as confrontation training and training under "complex electromagnetic environments" (CEME) (i.e., jamming, electronic warfare, cyber operations). Self-directed subjects are a relatively new concept in the PLAAF; they test pilots' skills under less scripted conditions and encourage more autonomy. For example, pilots at the squadron level are now being given the responsibility to create their own flight plans and have full autonomy over their sorties, from starting their engines to changing navigation routes and flying tactics in the air, to landing without strict control from a senior PLA Air Force commander in the control tower.[27] As discussed later in this chapter, this type of autonomy represents a significant departure from past practice, which emphasized strict ground control. Finally, "topics" are similar to "subjects" but appear to be of a higher degree of difficulty or feature sets of subjects that hone tactical skills. Pilots must pass in-flight exams on each training subject or topic before moving on to the next.

One development that has fundamentally changed how pilots assess performance is the use of information technology into the training process. For example, pilots are increasingly evaluated by senior flight leads using video feed playback (视频回放) and other flight-data recordings in "operations assessment and study rooms" (作战评估研究室). As noted in several recent *Kongjun Bao* articles, video playback offers important teaching tools to "rectify mistakes

---

[25] Wang Haihua, Zhao Lingyu, and Wang Caishan, 2013.

[26] *China Air Force Encyclopedia*, 2005, p. 277.

[27] Zhao Wen and Fan Jipeng, "Guangzhou MRAF Air Brigade Uses Informatized Means to Scientifically Organize Training," *Kongjun Bao*, March 27, 2012.

and deviations" (纠治偏差).[28] The use of video playback not only offers pilots news tools of evaluation but, more importantly, provides objective means of analysis for flight leads and instructors to hold debriefings. *Kongjun Bao* reporting notes the increasing prevalence of "calling out" experienced pilots for deviations from flight paths. This illustrates a shift in practice where perhaps in the past, flight leads might abstain from criticizing senior pilots.[29]

The *PLAAF Encyclopedia* refers to this daily training period on operational bases as the "consolidate and improve training stage" (巩固提高训练阶段), in which pilots focus on discrete tasks and build foundational aviation skills.[30] Historically, the goal of this phase of annual training has been to practice skills in a low-stress environment for application during the more complex and stressful "exercises" and "evaluations" during the summer and winter periods of the annual training cycle. However, the PLAAF has recently begun to add increasingly stressful and difficult daily training tasks into these periods. As one *Kongjun Bao* article states,

> In order to turn every pilot into a combatant "capable of fighting battles and winning battles," they put units in an "imminent battle-oriented" survival status at all times, set-up training subjects to be as difficult as possible, simulated a realistic battlefield environment, and organized different models of aircraft in subordinate units to carry out "back-to-back" confrontations under strong electromagnetic jamming and other such backgrounds.[31]

Aviation units have now begun experimenting with pairing veteran pilots with new pilots when flying low altitude over open ocean, which the PLAAF considers a "difficult" training subject. As the following *Kongjun Bao* passage indicates:

> There was a big difference between the levels of the technical foundations of each pilot, so the organization divided all the pilots participating in training into different levels and custom-set the training plans for each individual. They specially arranged for "one-on-one" paired-up mentoring with veteran pilots for the new pilots who were participating in the newly added subjects. Those who did not meet the requirements in mentored flight assessment absolutely were not allowed to conduct a solo flight in the related subject. The pilots who were participating in over-sea high-difficulty subject training for the first time also had to carefully view video data from past over-sea training and train on the contents of the related subjects in the flight simulator.[32]

---

[28] See, for example, Meng Qingbao, "Being 'Imminent Battle-Oriented' Becomes the Normal State of Training—A Certain Shenyang Military Region Air Force Air Division Sets Up a Vivid Battlefield Environment to Sharpen Air Combat Capabilities," *Kongjun Bao*, March 28, 2014.

[29] Meng Qingbao, 2014.

[30] *China Air Force Encyclopedia*, 2005, p. 274.

[31] Meng Qingbao, 2014.

[32] Yang Zhen and Xiao Qiangming, "Unidentified Regiment of Guangzhou Military Region Air Force Aviation Corps Scientifically Organizes Over-Sea High-Difficulty Subject Training," *Kongjun Bao,* May 28, 2013.

Peak Drill, Exercise, and Evaluation Season (June–August)

PLAAF pilots spend most of the summer months participating in exercises, drills, and evaluations. Exercises (演习) are multiday, major training events that involve real or simulated opposition force (对抗) and combined arms (合同作战) elements under competitive scenarios among either different branches within the PLAAF itself (aviation, surface-to-air missiles [SAMs], anti-aircraft artillery [AAA], and radar) or different services within the Chinese military as a whole (army, navy, Second Artillery). Exercises seek to integrate and apply the routine training subjects and skills practiced by aviators during the year. Based on articles from *Kongjun Bao*, there does not appear to be a major distinction between a training exercise (演习) and a training drill (演练). Both are "major" events taking place over one or more days and incorporate several training subjects that aim to mimic actual combat conditions and hone skills in combat and tactics.[33] Finally, evaluations (考验) appear to be smaller in scale than exercises and drills and assess discrete skills of pilots undertaking skills completion in combined arms settings.

According to *Kongjun Bao* reporting, most drills and exercises take the form of either "confrontational air battle assessments" (对抗空战考核) or "system-of-systems (SoS) confrontational drills" (体系对抗演练). For the former, only fighter aircraft and pilots participate in the test, but for the latter, multiple branches of the PLAAF are involved, including SAMs, radar, and AAA. Both evaluate pilots' ability to perform free-air combat (自由空战) and are meant to test pilot combat skills and tactics upon the removal of certain limits, such as decreasing the altitude gap between aircraft, and involve elements of "blue" versus "red" teams. However, for "confrontational air battle assessments," information about the engagement pattern, airspace, and altitude is prearranged and communicated to pilots in advance. SoS confrontational drills, in contrast, involve two confrontational parties that do not know about each other's conditions, and both parties have to rely on integrated information support, real-time communication, and self-judgment to achieve success. For this reason, the SoS confrontational drills involve a higher degree of uncertainty and complexity, as well as greater difficulty in detecting, identifying, and attacking targets. A typical "season" may feature anywhere from 6 to 12 drills and exercises involving both "confrontational air battle assessments" and "SoS confrontational drills."

---

[33] In theory, a training event is smaller in scale than a training exercise. However, several training events identified in the *Kongjun Bao* were comparatively large in scale, occurred over several days, and involved red and blue opposition force simulations with dissimilar aircraft.

A *Kongjun Bao* article of a SoS confrontational exercise at an unnamed airfield in 2013 provides a useful description of the types of elements involved.[34] According to the report, the drill started at dusk and involved one "red" and one "blue" team under "complex electromagnetic environments," using multiple-frame and "new-type fighter planes," including J-10s. There is no script for the exercise, as this passage indicates:

> The directing department does not define training subjects and other basic conditions; the combat objectives of the two belligerent parties, their respective operational plans, their use of the electromagnetic frequency spectral resources, and other specific details of the drill are not determined; the troops of both sides are not provided with intelligence and are instead expected to acquire intelligence mainly by their own means of reconnaissance.[35]

The red force can launch an attack any time after dawn, upon which the blue force can initiate countermeasures. Additionally, the directing department lifted restrictions on eight items, including, for the first time, choice of targets to attack, offensive-defensive countermoves, and timing of attack.

According to the sequence of events, starting around 0500, mobile radar, SAM, and other ground-support units participating in the drill maneuvered to their respective designated positions from each opposing team to prepare for combat. At 0524, the blue team fired signal flares and electronic jamming aircraft, and airborne early-warning aircraft took off, along with two J-10 combat aircraft. Right after that, airborne early-warning aircraft from the red team at a base several hundred kilometers away took off. This was the early warning and reconnaissance period of the exercise. At 0930, a combat alarm sounded from the blue team headquarters to intercept "enemy aircraft" approaching from a given heading. This initiated the "free air combat" period of the exercise. Pilots fought one another using various "unexpected offense and defensive tactics," including defense penetration at low altitude and surprise attack by outflanking the enemy in an attempt to seize the initiative. The red force reportedly staged a "very fierce and formidable" offensive, but the blue force had set up "multiple layers of defense," successfully repelling each successive red attack by adjusting its deployment and tactics. Both teams concluded activities and landed at dusk and proceeded to the post-exercise assessment room for evaluation. For this drill, there were reportedly no "winners" or "losers," just adherence to conditions of "actual combat."[36]

---

[34] Xu Tongxuan, "Another Drill in the Sky Above the Immense Desert, Closely Matching Actual Combat Conditions—An Eye-Witness Account of a 'Red-Versus-Blue' System-of-Systems Confrontational Drill at an Air Force Training Base," *Kongjun Bao,* October 14, 2013.

[35] Xu Tongxuan, 2013, p. 2.

[36] Xu Tongxuan, 2013.

Kongjun Bao reported on another exercise held in summer 2014. The relatively detailed description of tactics employed likely reflects the recent movement away from ground-controlled engagements to control by flight leads, as well as the evolution toward unscripted exercises.[37] The lead plane pilot of a red team squadron instructed his wingman (without apparent contribution from the ground controller), "Left turn 105 degree; target appears about 50 kilometers on your left front side; maintain assault formation; attack target number 2!" The wing plane made a rapid left turn, switched its radar to search for targets, and locked onto and successfully hit the target. Meanwhile, pilots from the blue team, who were under attack from the red team aircraft, dove, pulled up, and made rapid turns to avoid being targeted and hit. When the blue side was about to enter red's missile-attack range, the radar of the red team's wing plane was subjected to "powerful electromagnetic interference," completely "blacking out" the red team's monitor and losing the target. In response, the red team pilot released chaff and "conducted S shape abrupt turns to undermine the blue side's firing of missiles."

At the end of each exercise, pilots and ground crew from both red and blue sides sit side-by-side to participate in the evaluation and assessment period, which involves video playback, flight-data transmission feedback, and other electronic material.

Following the completion of the peak exercise season, units continue daily training in "Subjects" and "Topics" in the fall in preparation for Year-End Evaluations in the winter.

Year-End Evaluations (November–December)

Year-end evaluations (年终考核) are multiday exercises that typically occur in November and December. Evaluations are similar to the summer exercises in that they involve opposition-force conditions or high-difficulty training tasks and typically occur under extreme weather conditions. What makes evaluations unique is that pilots, aircraft, and training tasks are chosen at random to test training proficiency and readiness.

One *Kongjun Bao* article from November 2013, for example, highlighted the challenges faced by one particular Guangzhou MRAF air regiment that initiated year-end evaluations. According to the article, some of the aircraft identified to take part in the evaluations had just returned from "seasonal changeover maintenance" (换季工作), and four of the pilots chosen to participate were "ill-prepared" for examinations because they just had "just completed transition

---

[37] Sun Zhanrui and Chen Qingshun, "Conscious Wrestling and Close to Practice: A Certain Regiment of the Air Force of the Lanzhou Military Region Organizes Midair Confrontational Training in the Early Morning; Drill Director Only Issues Combat Operation Mission, Does Not Give Pre-Combat Operation Plans; Does Not Issue 'Notices on Enemy Situation'; Let the 'Red Army and Blue Army' Have 'Free Hands,'" *Kongjun Bao*, June 10, 2014, p. 2.

training" (完成改装训练不久).[38] For both challenges, the regiment leaders decided they would proceed with the evaluations as planned, justifying the risk by reminding the pilots that evaluations were meant to gauge "actual combat ability," not simply obtain "high marks." In both instances, the aircraft and pilots "successfully passed evaluations."[39]

Tactics and Combat Methods Training at the Operational Base

Woven into training tasks and evaluations are "tactics" and "combat methods." Tactics and combat methods training are held as elements within everyday training tasks as well as during exercises and evaluations within PLAAF aviation units.[40] According to the *PLAAF Encyclopedia*, tactics training comprises two parts: basic tactical flight training (战术飞行基础训练) and applied tactical flight training (战术飞行应用训练). Basic tactical flight training is designed to teach foundational knowledge and basic tactical maneuvers to flight personnel under rudimentary tactical circumstances.[41] This phase is the prerequisite for subsequent tactical flight applied training and uses tactical principles (战术原则) to integrate pilot skills with combat skills (战斗技术).[42] Applied tactical flight training is the second component of PLAAF tactical flight training and refers to training in which aviation units carry out comprehensive maneuvers to mimic what the PLAAF calls "realistic combat situations" (近似实战). It is the key phase of tactical flight training and builds on the foundation established during basic tactical foundation training. Applied tactics training involves coordinated operations training, opposition force training, and combined arms opposition force training.[43]

The passage below from a 2013 *Kongjun Bao* article summarizes some of the tactics certain PLAAF aviation units train in during a given training cycle:

> Unit training laid stress on such demanding tactical training tasks as penetrating enemy interception, assaulting tactical targets, and making penetrations against

---

[38] Luo Peiwu and Niu Haitao, "The Evaluation Site Is the Battlefield: Dispatches from a Guangzhou Military Region Air Force Air Regiment's Year-End Military Training Evaluations," *Kongjun Bao*, November 27, 2013.

[39] Luo Peiwu and Niu Haitao, 2013.

[40] The *China Air Force Encyclopedia* treats "combat methods" (战斗方法) as a component of "tactics" (战术). *Kongjun Bao* article also appears to treat the two concepts as interchangeable. In theory, however, tactics should represent the highest sophistication of aviation skills and reside above combat methods.

[41] *China Air Force Encyclopedia*, 2005, p. 295.

[42] *China Air Force Encyclopedia*, 2005, p. 295.

[43] For another assessment of PLAAF tactics training and combat methods, see Michael S. Chase, Kenneth W. Allen, and Benjamin S. Purser, III, *Overview of People's Liberation Army Air Force "Elite Pilots,"* Santa Monica, Calif.: RAND Corporation, RR-1416-AF, 2016.

strong jamming, flight above clouds, attack below clouds, maritime strike, long-range attack, electronic countermeasures, and combined arms operations. [44]

Estimates vary regarding the amount of time a typical Chinese fighter pilot devotes to tactics training at an operational base. One report from 2010, for example, estimates that at least 60 percent of all training at an operational unit is devoted to tactics training.[45] Other articles estimate that some units devote almost 80 percent of their "training plan schedule" (训练计划安排表) to tactical training.[46]

To illuminate the types or conditions and parameters associated with training activities held during a typical annual training cycle, Figure 4.4 provides data compiled on the top 12 types of training topics held across all PLAAF aviation units in 2013, shown as a percentage of total training activity for that year. Specific patterns vary by aircraft category and type, as do the nature of the exercises (with, for example, attack aircraft rehearsing ground attack and fighters largely holding air-to-air combat exercises). While broad in scope, the statistics give the reader a sense of the areas of emphasis tasked to PLAAF aviators for a given training cycle. PLAAF training often tests multiple training topics in a single mission, e.g., training at night and at low altitude.

---

[44] See Xu Weigang and Si Han, 2013.

[45] See National Air and Space Intelligence Center, 2010, p. 84.

[46] "Strictly Grasp Training Based on Standards—A Certain Air Aviation Regiment Actively Organizes Year-end Training Exercises" ["严格按规范抓训一级空航空某团积极稳定组织年底训练"] *Kongjun Bao*, December 23, 2013.

Figure 4.4. Training Topics Addressed in PLAAF Aviation Exercises (2013)

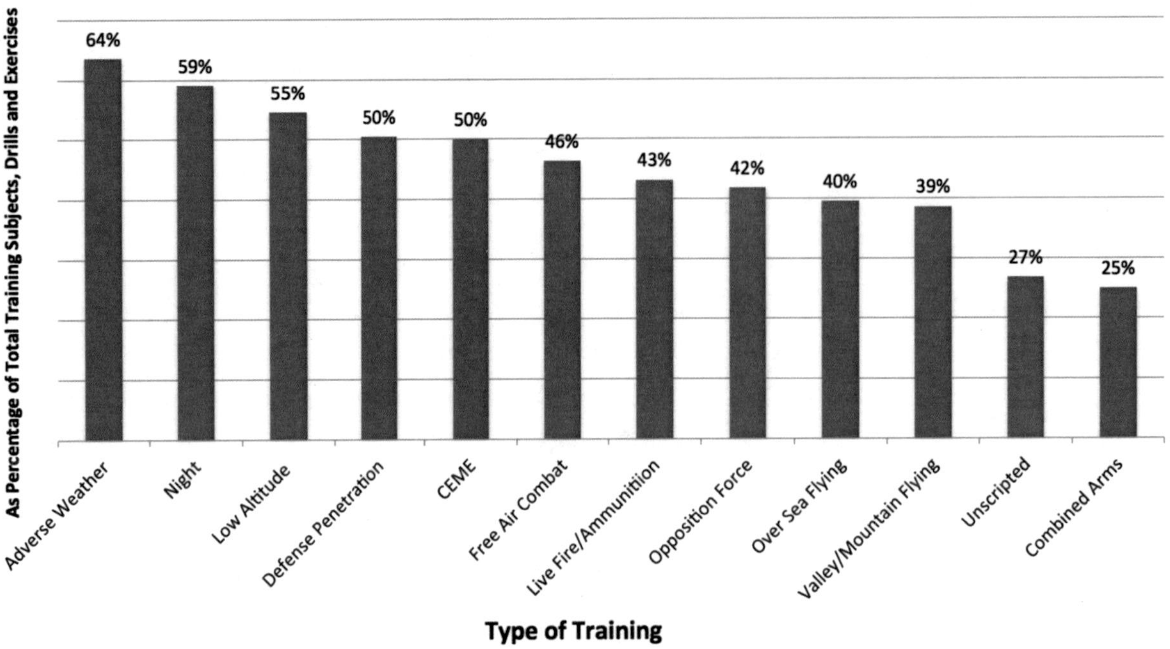

SOURCE: Compiled from all articles in *Kongjun Bao* in 2013. Based on total occurrences of training "subjects," "drills," and "exercises" identified for PLAAF aviation units for 2013, then coded for numbers of specific "types" of training activity that occurred as a percentage of the total. Note that pilots typically practice several different "types" of training within a given "subject," "drill," or "exercise."

5. PLAAF Aviation Instructor Training and Development

Of crucial importance in assessing the overall quality of PLAAF aviator development is the quality of the instructors at the Aviation University, flight colleges, and the units themselves. As one retired USAF fighter pilot wrote, "being the best pilot in the Air Force doesn't matter if no one can learn from you."[47]

The AWS is the primary mechanism for training USAF pilots to not only become instructors of airmen but also to hone elite weapons and combat skills. Based out of Nellis Air Force Base in Nevada, the school "teaches graduate-level instructor courses that provide the world's most advanced training in weapons and tactics employment."[48] The AWS works closely with the Nellis-based 422nd Test and Evaluation Squadron, which is also responsible for developing tactics. The AWS helps test and evaluate tactics developed by the 422nd, and its students disseminate those state-of-the-art tactics training to the rest of the USAF.

As outlined below, the PLAAF has demonstrated some interest in analogous systems, but has thus far developed only a rudimentary parallel. In developing new institutions to "train the trainers," the PLAAF has placed the focus primarily on raising the level of instructors at academies and flight schools, rather than on trainers engaged at the unit level.

For example, in the past, newly minted PLAAF cadets selected for instructor duty would typically stay at the flight college from which they graduated and self-study (自训) with one of the college's instructors.[49] They then would spend the rest of their career at the flight college. Very few pilots from operational units return to the flight colleges as flight instructors, opting instead to stay at the operational unit to move up the promotion ladder.[50] There is virtually no crossover from one career track to the other. As a result, the flight instructors at the academies lack sufficient technical and operational experience. One article likened this ad hoc style of

---

[47] Dan Hampton, *Viper Pilot,* New York: William Morrow, 2012, pp. 133.

[48] "U.S. Air Force Fact Sheet: United States Air Force Fighter Weapons School," May 10, 2016.

[49] "Certain Air Regiment of Air Force Flight Instructor Training Base Accelerates Training a New Type of Officer Training Corps" ["空军飞行教官训练基地某团加速培养新型教官队伍"], *China News Online*, September 25, 2013.

[50] For all pilots in an operational unit to move up the promotion ladder and eventually become a "special-grade" pilot, they must serve as a flight instructor for a certain number of hours per year. Whereas some pilots move up the ladder as "commanding officers" at each level (deputy commander and commander of flight squadrons, groups, regiments, brigades, and divisions), other pilots never serve in any type of a commanding role and can be a colonel flying for a captain.

training to "a pupil shadowing a master" (师傅带徒弟), gaining experience under the unstructured tutelage of individual instructors.[51]

The PLAAF seems to have recognized this shortfall. In April 2012, the PLAAF Flight Instructor Training Base (FITB) (空军飞行教官训练基地) was established as part of a major reform within the PLAAF aviation academy system to "institutionalize" and "professionalize" the training of aviation officers. The FITB, which can be thought of more as an officer training program than a school, is located in Bengbu, Anhui province, and was formed from the 13th Flight College and subordinated under the Air Force Aviation University.[52]

The opening ceremony of FITB was attended by PLAAF deputy commander He Weirong; deputy chief of staff Ma Zhenjun; president and political commissar of the PLAAF Air University Bai Chongming and Liu Dewei, respectively; and vice governor of Anhui province Huang Haisong, among others. At the ceremony, deputy commander He highlighted the need to "build a strong team of instructors and improve the core teaching ability and cultivate elite talent in line with the needs of a modern Air Force."[53] The attendance of high-ranking military officials at the ceremony indicates the seriousness with which the PLAAF regards the development of the FITB within the overall PLAAF training system.

One article in particular outlines in more detail the impetus and content behind the establishment of the program. According to the article, the aim is to create a "specialized training organ" (专门培训机构) to train a "new type of aviator [培养新型飞行人] able to command, understand combat, and manage" (会指挥, 懂作战, 善管理).[54] The program also serves to train pilots in the more advanced phases of combat and tactics training, including offensive and defensive mission sets; integrate learning across the different generations of airframes; and give flight instructors more operational experience.[55] In line with the concept of building standardized procedures and practices, the program has already adopted a series of teaching materials and officer handbooks. The program reportedly lasts six months, and pilots receive an accreditation

---

[51] "Certain Air Regiment of Air Force Flight Instructor Training Base Accelerates Training a New Type of Officer Training Corps," 2013.

[52] "PLA Air Force 13th Flight College" ["空军第十三飞行学院"], April 20, 2015; and "Air Force Flight Instructor Training Base Is Inaugurated" ["空军航空大学飞行教官训练基地在蚌揭牌"], *Bengbu News Online*, May 1, 2012.

[53] "Air Force Flight Instructor Training Base Is Inaugurated," May 1, 2012.

[54] "Certain Air Regiment of Air Force Flight Instructor Training Base Accelerates Training a New Type of Officer Training Corps," 2013.

[55] "Certain Air Regiment of Air Force Flight Instructor Training Base Accelerates Training a New Type of Officer Training Corps," 2013; and "Aviation Officer Training Says Goodbye to Inbreeding" ["飞行教官培养告别'近亲繁殖'"], *People's Daily Online*, April 7, 2014.

certificate (资格认证) when they complete the training program. The curriculum also benefited from the advice of "foreign instructors" (通过聘请外交).[56]

The limited amount of reporting on the program suggests that the 13th Flight College holds at least two different types of courses for different types of instructors. One 2013 article, which celebrated the graduation of the "first batch" of 20 pilots and eight transition training base pilots, noted that, after completing their six-month course of study, graduates were qualified to become flight instructors for new pilots.[57] They would not return to operational bases. Subsequently, however, a *People's Daily* article from April 2014 reported that the college had recently initiated a two-month "group training course" (集训) for pilots from operational bases across the country. As of spring 2014, the course had reportedly graduated 31 pilots.[58] The article lamented the "old" system of training instructors who remained at academies, having commanders at bases who stayed on as instructors at bases, and suggested that such a system had produced an "inbred" (近亲繁殖) form of training within the PLA. The new program being implemented aimed to "professionalize, systemize, and standardize" (走上专业化, 系统化, 规范化) training to offer aspiring instructors more hands-on experience learning sophisticated tactics and combat methods.[59] Thus, the course can be seen as an important development in creating linkages between operational units and PLAAF-wide educational institutions.[60] Although not stated directly, the article seemed to imply that the pilots return to their operational unit upon completion of the course.

On the surface, the six-month program at the FITB appears to have some parallels with the U.S. Air Force Pilot Instructor Training (PIT) program in the sense that it is designed to improve the skills of instructor pilots at the flight academies. However, unlike the PIT program, it is designed largely to give instructor pilots—who will never experience life or training at an operational unit—a sense of tactical aircraft handling. The two-month program, by contrast, appears designed to provide additional instruction for individuals chosen from among the best pilots at operational units who will then return to operational units as instructors. As such, it may

---

[56] "Certain Air Regiment of Air Force Flight Instructor Training Base Accelerates Training a New Type of Officer Training Corps," 2013. The article did not identify from which countries the "foreign instructors" came.

[57] "Certain Air Regiment of Air Force Flight Instructor Training Base Accelerates Training a New Type of Officer Training Corps," 2013. The following passage seems to indicate that graduates from the program stay behind at the flight academies to train incoming cadets: "在完成 6 个月的正规系统培训后，将取得具有资格认证的新型飞行教官，执教新一批空军飞行学员."

[58] "Aviation Officer Training Says Goodbye to Inbreeding," 2014.

[59] "Aviation Officer Training Says Goodbye to Inbreeding," 2014. The author used the term *unit-ize* (基地化) instructor training to describe this process.

[60] "Aviation Officer Training Says Goodbye to Inbreeding," 2014.

be seen as the closest thing that China has to the AWS. But a comparison of the two highlights the real point, which is that although the two-month FITB course could one day provide the genesis for a Chinese AWS, there simply is no parallel at the present time.

The AWS is highly selective and extremely competitive.[61] More important, it offers the highest levels of elite tactical and air-to-air and air-to-ground training in the world. Students progress through a grueling program of basic fighter maneuvers, air-combat maneuvering, air-combat tactics, and surface-attack tactics, to name a few, culminating in the mission-employment phase, which tests the students' ability to autonomously plan and execute the aforementioned skills and tactics in a campaign setting.[62] Graduates of the school then return to their operational units as experts in weapons and tactics with skills and knowledge they then impart to others in the unit. The two-month PLAAF FITB course, on the other hand, appears to focus primarily on flight theory and provides only limited opportunity for tactical training. Functionally, it is probably more analogous to the system of in-unit upgrade that sees U.S. pilots at the unit level polish their theoretical knowledge as well as flight skills.

The Chinese system for training instructors who will serve in flight academies and the system for developing flight instructors at the unit level suffer from the strict stove-piping of career paths within the PLAAF. The system as a whole suffers the lack of a school that integrates lessons learned at advanced tactical centers with the development of unit-level instructors. Nevertheless, the FITB is an important new institution with the possibility of developing further to address some of the gaps in the Chinese pilot-training system.

---

[61] According to Dan Hampton, only 30 fighter pilots are selected from hundreds across the United States and abroad, including only three to four F-16 pilots. See Hampton, 2012, p. 133.

[62] For an in-depth account of this experience, see Hampton, 2012, pp. 132–140.

6. Assessing Progress and Weaknesses in PLAAF Aviation Unit Training

Progress

Progress in this case is measured by the PLAAF's stated criteria of training and equipping PLAAF pilots with the skills necessary to "fight and win battles," with an emphasis on holding training under "actual combat" (实战化) scenarios. By these measures, the PLAAF is clearly moving toward less-scripted scenarios to cultivate the aviator's ability to think and react more autonomously and under more complex environmental and technological conditions.

Perhaps the most prominent components of PLAAF unit training in recent years are the dual concepts of "free air combat" (自由空战) and "unscripted scenarios" (不知条件下). Placing the focus on these two concepts illustrates the PLAAF Party Committee's emphasis on executing the broad directives for the PLA as a whole to train under "actual combat" conditions. *Kongjun Bao* reports suggest that PLAAF unit exercises and drills frequently remove safety restrictions—such as the distance between opposing aircraft—or refrain from sharing the preflight plans of opposing forces in an effort to simulate unscripted scenarios. The ultimate goal of these drills is to cultivate autonomous decisionmaking among pilots, ground crew, and commanding officers when presented with the uncertain characteristics of war.

The following quote, provided in a *Kongjun Bao* report from a August 2012 exercise, suggests that the transition to "free air combat" is a work in progress, with significant interaction between ground controllers and flight leads, but a somewhat higher expectation that the latter will at least make final tactical decisions:

> The flight crew of the division that were undertaking the anti-ground assault task were conducting ultra-low altitude concealed penetration. Suddenly, the aircraft was tracked by the "enemy's" new-type radar. The crew immediately commenced jamming signals, sped up toward the right side, and reported to the command and control center. The ground commander led the command team to immediately activate the flight task planning system, reselect the penetration route, and send it to the handheld terminals carried by the crew. In light of the threat source detection and attack blind zone, Crew Commander Ji Wanchao independently made the decision to launch a 100-meter ultra-low altitude concealed penetration.[63]

---

[63] An Hongxin, Li Guan, and Cao Chuanbiao, "Focus: Quickly Generate Combat Power Along Pace of Times—Record of Unidentified Lanzhou Military Region Air Force Bomber Air Division Using Information Support Systems to Accelerate Transformation of Flight Organization and Command Model," *Kongjun Bao*, August 15, 2012.

Another indication of progress in PLAAF training is the increasing use of video playback and more sophisticated data-transmission feeds, which appear to allow for more honest and transparent assessments of aviators. This has improved the mechanisms by which PLAAF pilots can learn from their mistakes. One report notes how such tools are exposing the flaws of veteran pilots:

> In the process of aerobatic flight, when a certain pilot changed from a vertical maneuver, there was a deviation of more than 10 degrees. Deputy Regiment Commander Xu, who was responsible for the evaluation, started right off the bat by criticizing two veteran pilots by name for the problem of not maintaining flight data strictly enough, and every word of his criticisms rang true and cut deep. This regiment took the attitude of "zero tolerance" for personnel and situations that had problems in flight training. No matter whether it was a leading cadre or an ordinary pilot, they always called them out by name and spoke bluntly in summation and evaluation.[64]

Similar reports highlight a shift in the way pilots are evaluated, from a focus on simply achieving high marks under scripted scenarios to a focus on the issues encountered under conditions of "actual combat":

> There should be nothing to fear about more vigorously exposing existing problems rather than achieving superficial results in combat operations that appear to be brilliantly conducted and divorced from actual combat. It would be much better to achieve slightly poorer performance scores in an assessment that adheres closely to actual combat.[65]

Another report notes, for example, the importance of seeing "their own mistakes from their opponent's perspective," in order to "improve and rectify their own weaknesses before the enemy does."[66]

Furthermore, the PLAAF appears to be improving its ability to routinize training on dissimilar aircraft and training pilots of those aircraft at a single air base or operational unit. One *Kongjun Bao* article reports how the PLAAF is adopting innovative approaches to training pilots of second- and third-generation aircraft (the Chinese terminology for third- and fourth-generation aircraft):

---

[64] Meng Qingbao, "The Sharp Edge of the Blade Comes from Fine Polishing—Record of Personal Experience in Implementing Strict Administration of Training Throughout the Whole Course of Flight Training by a Certain Regiment of the Shenyang Military Region Air Force Aviation Corps," *Kongjun Bao*, April 17, 2013.

[65] Lou Yongjun and Liu Daquan, "An Eye-Witness Account of the Self-Criticisms and Self-Reflections Made by a Shengyang Military Region Air Force Air Regiment About Its Participation in the PLA Competition-Oriented Assessment of Defense-Penetration and Assault Capabilities," *Kongjun Bao*, July 10, 2014.

[66] Xin Jianjun and An Hongxin, "Gazing at the 'Mirror' of Combat Realism, Finding Deficiencies in Training: Lanzhou MRAF Brigade Strengthens Problem-Driven Approach, Breaks Through Training Difficulties," *Kongjun Bao*, April 25, 2014.

The division broke away from the practice of transferring pilots up from units equipped with second-generation aircraft to units equipped with third-generation aircraft by transferring five groups of outstanding backbone third-generation aircraft pilots, totaling 10 pilots, to posts in units equipped with second-generation aircraft with the aim of nurturing flying personnel who were competent at manning both the second-generation and third-generation aircraft in combat operations and at commanding other pilots. In so doing, the division facilitated the overall elevation of its flying personnel combat capability.[67] [Note: the generations mentioned here are stipulated using the Chinese definition and translate into U.S. third and fourth generation.]

The efficient integration of different generations of aircraft within the PLAAF remains a work in progress, highlighted by the generation gap between veteran pilots and pilots of newer-generation aircraft. By allowing third-generation pilots to train on second-generation aircraft, it appears the PLAAF is attempting to cultivate an aviator corps equipped with knowledge to fly both generations of aircraft.

Finally, pilots are increasingly being deployed to units and airfields other than their home unit in an effort to hone skills in reorienting to unfamiliar locations. As the following excerpt from *Kongjun Bao* states, the practice presents challenges for pilots who are unfamiliar with adapting to a new environment:

> Right after formulating combat power using newly equipped aircraft, this air force evaluation required these four air crews to take off from an unfamiliar highland airstrip, conduct super-low-altitude penetration along an unfamiliar course, and travel several thousand kilometers to carry out a live ammunition bombing of a real and unfamiliar target for the first time, which was a completely new challenge. They specially set up several problem tackling teams led by regiment leaders that focused on such topics as bomb dropping, navigation, and operational methods.[68]

Shortcomings

Because of biases in Chinese-language military newspaper reporting, it is sometimes difficult to accurately assess the weaknesses in PLAAF pilot training. Most *Kongjun Bao* reporting, for example, overstates successes and underreports failures or major flaws in execution. However, some reporting alludes to shortfalls in training execution that are nonetheless discernable from Chinese reporting. Shortfalls include insufficient flight-lead skills and autonomy, lax discipline

---

[67] Sun Lin and Yang Jin, "Backbone Flying Personnel of Third-Generation Aircraft Help Nurture the Pilots of Units Equipped with Second-Generation Aircraft" ["三代机骨干反哺二代机部队"], *Kongjun Bao*, July 17, 2014.

[68] Tan Lei and Chi Yuguang, "Strictly Studying Detailed Training, Strengthening Stamina for Winning—A Certain Nanjing Military Region Air Force Air Regiment Firmly Establishes Combat Power Standards in Making Up for Shortcomings and Weaknesses," *Kongjun Bao*, July 9, 2014.

during daily training routines, underdeveloped tactics, and lack of coordination with other PLAAF branches.

The role of flight leads is crucial in a unit's ability to carry out combat missions. Flight leads not only guide wingmen during operations, they possess adequate experience in tactical maneuvering and autonomous decisionmaking that is required under actual combat scenarios. As the following passage makes clear, PLAAF units continue to lack flight leads with sufficient tactical skills:

> A prominent problem was that the tactical understanding and application of formation lead planes did not converge with the situations. It can be said that all the lead planes were generally technically oriented lead planes that handed over all "air combat" command power to ground command and guidance personnel during confrontations. As such, there were many unfavorable factors that come about during air combat. For example, ground commands often are not able to keep up with the complex and changeable air situation. Pilots relied too much on the commands and guidance from the ground, which was not conducive to enhancing the enthusiasm and initiative of airborne combatants. Therefore, it is necessary to have lead planes develop "tactical skills" over "technical skills" and take on more of a lead role without delay.[69]

As another passage makes clear, handing over tactical operations from the ground control to flight leads remains a work in progress for the PLAAF:

> We should re-divide the work of commanding and hand over the power over tactical actions that were decided by the command post in the past to the lead plane, so that the command post only has to inform the pilot about the air situations and the lead plane can assume full command over tactical actions, with the command post only performing the necessary monitoring. Specifically, the command post can provide guidance for mid and long range targeting, and at a distance of xx km within the target, it should hand over the command power to the lead plane, so that the lead plane can command the wingmen or the wingmen can take tactical actions.[70]

Although the PLAAF has prioritized honing pilot skills under unscripted scenarios, the concept remains new to many pilots accustomed to having most, if not all, of their tactical maneuvers dictated to them by PLAAF unit commanders in the control tower:

> When making preparations (for the low-altitude defense penetration training task), some pilots set up various contingencies beforehand. But since targets were changed on an ad hoc basis, they ended up being too close for accurate aim and the pilots thus failed to shape a launch condition. Some pilots did not appropriately study tactics and combat methods and only paid attention to narrow requirement of penetration at an ultra-low altitude. During actual ground assault

---

[69] Cheng Yongfeng, "Aviation Theory Classroom: How to Turn 'Technical Skills' of Formation Lead Planes into 'Tactical Skills,'" *Kongjun Bao,* April 9, 2014.

[70] Cheng Yongfeng, 2014.

time, they became too nervous because they did not appropriately understand and distribute tactical knowledge. In the end they failed to find the target or hit a wrong target.[71]

Furthermore, some reporting alludes to complacency in daily training, even among veteran pilots:

> Why do veteran pilots who have thousands of flight hours under their belts and top flight talents who have earned great honors in major military tasks have "slip-ups" in simple subjects in everyday training? Regiment Commander Gu's words resonated with everyone: "When executing major exercises, everyone's focus is highly concentrated and no one would dare to be negligent in the slightest degree. However, some pilots see regular training as "ordinary flying," so they intentionally or unintentionally lower their requirements and standards.[72]

Finally, the process of planning and executing large multiday exercises, as examined earlier in the chapter, has uncovered issues regarding insufficient coordination between aviation units and the different branches of the PLAAF, namely the surface-to-air forces:

> When the evaluation rules (for exercises) are revised in the future, more land-based missile units should be arranged to fight along with, and not against, the test-participating aircraft units from the same military regions; in the meantime, the plane-catching signals of the missile battalions should be transmitted in a real-time mode to the command hall. This may let the ground and air forces promote each other's capabilities.[73]

Based on the language in the above passage, it appears that SAM units are frequently employed as opposing forces against aviation units during multibranch exercises, rather than working in tandem with aviation units against an opposing force.[74] This distinction has implications for joint and combined arms training, which places a premium on creating interoperational communication networks among the various PLAAF branches. As the following self-assessment from another exercise makes clear, PLAAF officials have identified a need to improve "mutual critiques" involving different opposing sides, to include air defense and aviation units:

> We had an opportunity to read a self-assessment report by a ground-based air defense unit for the first day of the drill, which not only provided a summary of

---

[71] Xu Tongxuan and Li Wuchao, "Charge Toward Combat-Realistic Training—Account of the Penetration and Assault Tests with a Competitive Feature Carried Out by the PLA Air Force for the Aviation Units in 2014," *Kongjun Bao,* June 26, 2014.

[72] Meng Qingbao, 2013.

[73] Xu Tongxuan and Li Wuchao, 2014, p. 1.

[74] Although Lin and Garafola find that PLAAF SAM units train in both supporting and opposing force roles within combined arms training, they reach a similar conclusion regarding the use of SAMs more often than not in opposing force roles. See Lin and Garafola, 2016.

the unit's existing problems but also pointed out its opponents' faults. A staff officer with the rank of captain serving the assessment group noted: If it was not due to restriction of location and if the two sides could sit together to hold a mutual critique seminar at the end of every day of the drill, then the two sides would definitely be able to learn from each other better and help each other forward.[75]

---

[75] Xu Tongxuan, 2013, p. 3.

7. Overall Assessment

The PLAAF has been taking elements of actual combat (实战) subjects, topics, exercises, and methods as the basis for transforming its military training system and improving its level of operational effectiveness. Although themes associated with "actual combat"[76] have been featured in key strategic training guidance documents for years within the PLAAF and Chinese armed forces as a whole, it appears senior PLAAF leaders have redoubled their efforts at instilling discipline and offering honest assessments of shortcomings across all levels of aviation unit training.[77] Recently created senior decisionmaking mechanisms—such as a "Leading Small Group on Military Training Monitoring" (全军军事训练监察领导小组)—serve to reinforce bureaucratic oversight over the process.[78]

Assessments of training competence are of course subject to the criteria used as the basis of analysis. Based on the types of training tasks being reported in *Kongjun Bao*, there has been a clear increase in the degree of difficulty of training subjects and scenarios that have been stipulated by the PLAAF Party Committee. These include flying under challenging environmental conditions, such as at night and extreme weather patterns; flying at low altitudes through valleys and mountains and over water; cultivating "free air combat" skills among aviators with decreased altitude restrictions; and holding sophisticated multibranch and service exercises under complex electromagnetic environments and formidable air defense scenarios to mimic actual battle conditions that a potential military adversary may present. In a significant shift from prior practice, pilots in some units are now given the responsibility to create their own flight plans and have full autonomy over their sorties with little guidance from ground control.

There remain significant barriers to development as well. Institutional impediments run deep in a military that has for decades remained an army-centric fighting force. The mission sets that the PLAAF has been asked to train for in support of broader Chinese strategic objectives remain underdeveloped. As a result, concepts involving joint command and control across different branches within the PLAAF itself as well as across the army, navy, and air force are just gaining currency. When compared with its USAF counterpart, clear deficiencies remain among PLAAF pilots in the area of combat tactics and skills. Despite the PLAAF's emphasis on "free air

---

[76] References to "实战" have appeared in every Chinese Defense white paper since 2000, for example.

[77] A corollary to the PLAAF's effort to institute more rigorous standards of pilot training proficiency can also be found in China's effort to overhaul its civilian airline pilot training system. See, for example, James Fallows, *China Airborne: The Test of China's Future*, 1st ed., New York: Vintage Books, 2013.

[78] Liang Pengfei, "Military-Wide Military Training Monitoring Work Will Continue and Broaden," *Jiefangjun Bao*, January 24, 2015.

combat," PLAAF pilots appear to fall short on the requirement of executing autonomous actions or sophisticated aerial maneuvers in unscripted environments. As with any reform process, the path to building a modern fighting military force remains a work in progress.

Appendix. Notional USAF Versus PLAAF Fighter Pilot Pre-Unit Training

U.S. Air Force					People's Liberation Army Air Force				
PME Phase	Location	Time	Content	Aircraft	PME Phase	Location	Time	Content	Aircraft
Academics									
Undergraduate education; Officer Training School (OTS)	Air Force Academy; OTS	9 weeks–4 years	Basic undergraduate military education, physical fitness, leadership, and drills	N/A	PLAAF Aviation University, phase 1*	PLAAF AU	3.5 years	Basic academics, military doctrine, aviation theory, psychology, parachuting, and survival training	N/A
Basic Flight Training									
Undergraduate Pilot Training (UPT), phase 1	Air Force Base	6 months	Ground operations, take-offs and landings, spins, aerobatics, and emergency procedures	T-6 Texan	Basic flight training, phase 1	PLAAF AU	6 months	Take-offs, landings, spins, emergency procedures, basic navigation, instrument flying, night flying, and formation flying	CJ-6
Advanced Flight Training									
UPT, phase 2	Air Force Base	6 months	Aerodynamics, aircraft systems, weather, and instrument flying procedures	T-38 Talon	Professional aviation education, phase 2	One of three PLAAF flight colleges	1–2 years	1) ground-based training, including sitting in a cockpit and simulator training; 2) flying with a flight instructor; and 3) solo flight training; content includes takeoff and landing routes, airspace, instruments, formation flying, simple aerobatics, and navigation	K-8; L-15; JJ-7; Q-5; Y-7; Z-9
Introduction to Fighter Fundamentals (IFF)/ Formal Training Unit (FTU)	Air Force Base	3 months for IFF/ 7–9 months for FTU	Dropping bombs, strafing, dogfighting, centrifuge training, self-protection systems, bad-weather flying, mid-air refueling, SAM defense	T-38/follow-on fighter (F-16, F-15, F-22, A-10)	Combat transition training, phase 3	Separate transition training unit within an air force base	1 year	Four-weather training (e.g., day, night, different weather conditions), training in combat subjects, campaign and tactics training, and combined arms training	Advanced supersonic trainer, operational unit aircraft

SOURCES: Reporting from *Kongjun Bao*; U.S. Air Force, "Operations Training Syllabi, Air Combat Command," 2013–2015; Hampton, 2013; National Air and Space Intelligence Center, 2010.

References

"Air Force Flight Instructor Training Base Is Inaugurated" ["空军航空大学飞行教官训练基地在蚌揭牌"], *Bengbu News Online*, May 1, 2012. As of June 6, 2016: http://www.bbnews.cn/xw/bb0yw/2012/05/383611.shtml

Allen, Kenneth W., "PLA Air Force, Naval Aviation and Army Aviation Aviator Recruitment, Education, and Training," Washington, D.C.: The Jamestown Foundation, February 2016.

An Hongxin, Li Guan, and Cao Chuanbiao, "Focus: Quickly Generate Combat Power Along Pace of Times—Record of Unidentified Lanzhou Military Region Air Force Bomber Air Division Using Information Support Systems to Accelerate Transformation of Flight Organization and Command Model," *Kongjun Bao*, August 15, 2012.

"Aviation Academies Use Supersonic Fighter in Pilot Training" ["超音速歼击机首次进入空军飞行院校"], *China Military Online*, February 28, 2015. As of April 22, 2015: http://english.chinamil.com.cn/news-channels/2015-02/28/content_6372467.htm

"Aviation Officer Training Says Goodbye to Inbreeding" ["飞行教官培养告别'近亲繁殖'"], *People's Daily Online*, April 7, 2014. As of May 23, 2016: http://www.gywb.cn/content/2014-04/07/content_578987.htm

"Certain Air Regiment of Air Force Flight Instructor Training Base Accelerates Training a New Type of Officer Training Corps" ["空军飞行教官训练基地某团加速培养新型教官队伍"], *China News Online*, September 25, 2013. As of May 23, 2016: http://www.chinanews.com/mil/2013/09-25/5321140.shtml

Chase, Michael S., Kenneth W. Allen, and Benjamin S. Purser III, *Overview of People's Liberation Army Air Force "Elite Pilots,"* Santa Monica, Calif.: RAND Corporation, RR-1416-AF, 2016. As of August 2016: http://www.rand.org/pubs/research_reports/RR1416.html

Cheng Yongfeng, "Aviation Theory Classroom: How to Turn 'Technical Skills' of Formation Lead Planes into 'Tactical Skills,'" *Kongjun Bao,* April 9, 2014.

Cliff, Roger, John F. Fei, Jeff Hagen, Elizabeth Hague, Eric Heginbotham, and John Stillion, *Shaking the Heavens and Splitting the Earth: Chinese Air Force Employment Concepts in the 21st Century*, Santa Monica, Calif.: RAND Corporation, MG-915-AF, 2010. As of May 23, 2016: http://www.rand.org/pubs/monographs/MG915.html

China Air Force Encyclopedia, Vol. 1, Yao Wei, ed., Beijing, China: Aviation Industry Press, November 2005.

Fallows, James, *China Airborne: The Test of China's Future*, 1st ed., New York: Vintage Books, 2013.

Garamone, Jim, "Pace Visits Chinese Air Base, Checks Out Su-27 Fighter-Bomber," *American Forces Press Service*, March 24, 2007.

Hallion, Richard P., Roger Cliff, and Phillip C. Saunders, *The Chinese Air Force: Evolving Concepts, Roles, and Capabilities*, Fort McNair, Washington, D.C.: National Defense University, 2012.

Hampton, Dan, *Viper Pilot,* New York: William Morrow, 2012.

Liang Pengfei, "Military-Wide Military Training Monitoring Work Will Continue and Broaden," *Jiefangjun Bao*, January 24, 2015.

———, "PLA General Staff Department Promulgates Directive on 2015 PLA Military Training," ["总参部署新年度全军军事训练任务"], *Jiefangjun Bao*, January 18, 2015.

Liang Pengfei, and Li Yuming, "Joint Examination on Trial Training Outline of All-Army New Training Program Completed," *China Military Online*, February 5, 2015.

Lin, Bonny, and Cristina L. Garafola, *Training the People's Liberation Army Air Force Surface-to-Air Missile (SAM) Forces*, Santa Monica, Calif.: RAND Corporation, RR-1414-AF, 2016. As of September 2016:
http://www.rand.org/pubs/research_reports/RR1414.html

Liu Fengan, and Wu Tianmin, "New-Generation 'Outline of Military Training and Evaluation' Promulgated" ["新一代《军事训练与考核大纲》颁发"], *Jiefangjun Bao*, July 25, 2008, p. 1.

Lou Yongjun, and Liu Daquan, "An Eye-Witness Account of the Self-Criticisms and Self-Reflections Made by a Shengyang Military Region Air Force Air Regiment About Its Participation in the PLA Competition-Oriented Assessment of Defense-Penetration and Assault Capabilities," *Kongjun Bao*, July 10, 2014.

Luo Peiwu, and Niu Haitao, "The Evaluation Site Is the Battlefield: Dispatches from a Guangzhou Military Region Air Force Air Regiment's Year-End Military Training Evaluations," *Kongjun Bao*, November 27, 2013.

Meng Qingbao, "Being 'Imminent Battle-Oriented' Becomes the Normal State of Training—A Certain Shenyang Military Region Air Force Air Division Sets Up a Vivid Battlefield Environment to Sharpen Air Combat Capabilities," *Kongjun Bao*, March 28, 2014.

National Air and Space Intelligence Center, *People's Liberation Army Air Force 2010*, Wright-Patterson Air Force Base, Ohio, National Air and Space Intelligence Center: August 1, 2010.

People's Liberation Army Military Training Regulations (中国人民解放军军事训练条例), Beijing, China: PRC Central Military Commission (中央军委), April 6, 1990. As of April 27, 2015:
http://www.mod.gov.cn/policy/2009-07/14/content_3100976.htm

"PLA Air Force 13th Flight College" [空军第十三飞行学院], April 20, 2015. As of April 20, 2015:
http://baike.baidu.com/view/743305.htm

"Strictly Grasp Training Based on Standards—A Certain Air Aviation Regiment Actively Organizes Year-End Training Exercises" ["严格按规范抓训一级空航空某团积极稳定组织年底训练"], *Kongjun Bao*, December 23, 2013.

Sun Lin, and Yang Jin, "Backbone Flying Personnel of Third-Generation Aircraft Help Nurture the Pilots of Units Equipped with Second-Generation Aircraft" ["三代机骨干反哺二代机部队"], *Kongjun Bao*, July 17, 2014.

Sun Zhanrui, and Chen Qingshun, "Conscious Wrestling and Close to Practice: A Certain Regiment of the Air Force of the Lanzhou Military Region Organizes Midair Confrontational Training in the Early Morning; Drill Director Only Issues Combat Operation Mission, Does Not Give Pre-Combat Operation Plans; Does Not Issue 'Notices on Enemy Situation'; Let the 'Red Army and Blue Army' Have 'Free Hands,'" *Kongjun Bao*, June 10, 2014, p. 2.

Tan Lei, and Chi Yuguang, "Strictly Studying Detailed Training, Strengthening Stamina for Winning—A Certain Nanjing Military Region Air Force Air Regiment Firmly Establishes Combat Power Standards in Making Up for Shortcomings and Weaknesses," *Kongjun Bao*, July 9, 2014.

Tian Wei, "Solidly Do a Good Job in Conducting Flight Training in the New Annual Cycle with Concentrated Attention and Rigorous Organization" ["集中经历严密组织切实抓好新年度飞行训练"], *Kongjun Bao*, January 8, 2014.

U.S. Air Force, "Operations Training Syllabi, Air Combat Command," 2013–2015.

"U.S. Air Force Fact Sheet: United States Air Force Fighter Weapons School," May 10, 2016. As of May 23, 2016:
http://www.nellis.af.mil/AboutUs/FactSheets/Display/tabid/6485/Article/284156/united-states-air-force-weapons-school.aspx

Wang Haihua, Zhao Lingyu, and Wang Caishan, "'Airborne Hammer' Strikes Like a Thunderbolt—Eyewitness Account of Strict and Assiduous Organization of Initial Flight

Training at Start of the New Year While Closely Adhering to Requirements of Actual Combat by a Certain Regiment of the Guangzhou Military Region Air Force Aviation Corps," *Kongjun Bao,* January 9, 2013.

Wang Zhonghai, and Liu Daquan, "Air Regiment of the Shenyang Military Region Air Force Energetically Conducts Core Capabilities Training, Hones Defense Penetration and Assault Skills," *Kongjun Bao*, February 10, 2014.

Xin Jianjun, and An Hongxin, "Gazing at the 'Mirror' of Combat Realism, Finding Deficiencies in Training: Lanzhou MRAF Brigade Strengthens Problem-Driven Approach, Breaks Through Training Difficulties," *Kongjun Bao*, April 25, 2014.

Xu Tongxuan, "Another Drill in the Sky Above the Immense Desert, Closely Matching Actual Combat Conditions—An Eye-Witness Account of a 'Red-Versus-Blue' System-of-Systems Confrontational Drill at an Air Force Training Base," *Kongjun Bao,* October 14, 2013.

Xu Tongxuan, and Li Wuchao, "Charge Toward Combat-Realistic Training—Account of the Penetration and Assault Tests with a Competitive Feature Carried Out by the PLA Air Force for the Aviation Units in 2014," *Kongjun Bao,* June 26, 2014.

Xu Weigang, and Si Han, "An Air Brigade of the Nanjing Military Region Air Force Organizes Highly Demanding Training in Accordance with Actual Operation Requirements" ["紧贴实战应用组织高难课目训练"], *Kongjun Bao,* June 27, 2013.

Yan Guoyou, and Liao Qirong, "PLA Nanjing MRAF Division: Combat Realistic Training Routinized" ["南空航空兵某师常态化开展实战化训练"], *Kongjun Bao,* August 15, 2013.

Yang Zhen, and Xiao Qiangming, "Unidentified Regiment of Guangzhou Military Region Air Force Aviation Corps Scientifically Organizes Over-Sea High-Difficulty Subject Training," *Kongjun Bao,* May 28, 2013.

Zhao Wen, and Fan Jipeng, "Guangzhou MRAF Air Brigade Uses Informatized Means to Scientifically Organize Training," *Kongjun Bao*, March 27, 2012.